COMPETITIVE PHYSICS 2

INTRODUCTION

This objective physics series provides a basic and challenging problem of physics from particular topics. It can be used to brush up ones basics and checking up the preparation level of particular topics. It is equally helpful to the traditional classes as well as competitions. It can be also taken as a revision material for any competition which includes the test of basic physics. If you want to grasp the subject before practicing these multiple choice questions, you can go through the website http://www.ncert.nic.in/ncerts/textbook/textbook.htm and down load the free copy of science books and after having command on the topic practice it. For revision purpose, important points are given at the starting of each topic.

CONTENTS

LIGHT

SOME IMPORTANT POINTS

- LAWS OF REFLECTION:
  1. Angle of incidence is equal to angle of reflection
  2. The incident ray, reflection rays the normal at the point of incident all lie in same plane.
- Concave mirror are converging mirror convex mirror are diverging mirror.
- The center of reflecting surface of spherical mirror is pole.
- A imaginary line passing through the pole and centre of curvature of a spherical mirror is called principal axis.
- The distance between the pole and principal focus of the spherical mirror is called focal length it is denoted by f.
- The relationship between the radius of curvature and focal length spherical mirror is shows by formula F=R/2.
- In convex mirror image is formed virtual and diminished.
- In concave object placed at between P and F image is virtual enlarged.
- Concave mirror ate used in headlight of vehicles in field of solar energy in barber shop etc.
- Convex mirror are used in rear view of traffic in vehicles and in search mirror in big shops.
- Mirror formula :1/v+1/u=1/f
- The ratio height of image and height of object is known as magnification of mirror

  $M = h_i/h_o = -v/u$ [for mirror]

- The change in direction of light when it passes from one transparent medium to another is called refraction of light.
- Laws of refraction: The incident ray refracted ray and normal to surface of separation of two transparent media at point of incidence all lie in some plane.

- The ratio of sine of angle of incidence to the sins of angle of refraction is constant for a given pair of medium and colour.
- The value of sin i/sin r for a ray of right passing through one medium to ant there is known as refraction index.
- Spherical lens are types:
- Concave lens and convex lens.
- Concave lens are diverging lens and convex lens are converging lens.
- In concave lens image formed is virtual and diminished.
- In convex lens image formed is real, diminished, enlarged and same size.
- Lens formula :1/v-1/u=1/f
- Magnification of lens : m=h'/h=v/u

  h'=image height ,h=object height , v=image distance ,u=object distance
- The reciprocal of focal length is known power of lens. Its S.I. unit is Dioptre and denoted by D.
- Power of concave lens is negative and convex lens has power positive.

6. LIGHT

1.   The image formed by a plane mirror is always

      (a)   Real and erect   (b) Virtual and inverted

      (c)   Real and inverted (d)   virtual and erect

2.   Which mirror always gives virtual and erect image?

      (a) Concave mirror (b) Convex mirror (c) Plane mirror   (d) both (b) & (c)

3.   The reflecting surface of a mirror curved inward is known as:

      (a) Concave (b) Convex   (c) Plane   (d) both (a) & (b)

4.   The centre of reflecting surface of a spherical mirror is known as:

      (a) Pole   (b) Polo   (c) Centriod   (d) point

5.   The centre of Curvature of Concave mirror lies:

      (a) In backward   (b) In front of its reflecting surface

      (c) On the mirror   (d) none of these

6.   The relationship between P and f of a spherical mirror is :

      (a) P=F   (b) P=2F   (c) P=2f   (d) p=f/2

7.   The image formed by a concave mirror is always

      (a) Real and erect (b) virtual and inverted

      (c) Real and inverted   (d) None of these

8.   Which type of mirror forms a real and same size image of the object:

      (a) Plane   (b) Concave (c) Convex   (d) both (a) & (c)

9.

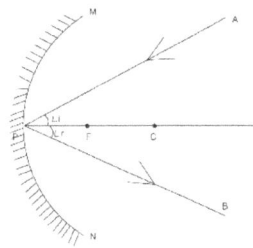

In the figure $\angle i = 60^0$, then find $\angle BPN=?$

(a) 60^0 (b) 30^0 (c) 90^0 (d) 45^0

10.   The focal length of a mirror is 29.5m then find its radius of curvature?

        (a) 29.5   (b) 59   (c) 59.5   (d) None of these

11.   Which mirror is used as a rear view mirror in vehicles?

        (a) Concave   (b) Plane   (c) Convex   (d) All of the above

12.   The focal length of a mirror is 28cm, object distance is 15cm then find the image distance.

        (a) 420/13 m   (b) 420/13 cm   (c) 13/420 cm   (d) none of these

13.   The magnification of an object is -0.5, then find the nature and size of the image

        (a) Real and enlarged   (b) virtual and diminished

        (c) Real and Diminished   (d) both (a) & (c)

14.   The height of the image is 0.15, and then finds the nature of the image

        (a) Real and inverted   (b) virtual and erect

        (c) Real and erect   (d) virtual and inverted

15.   The object distance is always:

        (a) Positive   (b) negative   (c) depend on the image   (d) Depend on the mirror

16.   The distance between the pole and centre of the mirror is known as:

(a) PC   (b) Radius of curvature   (c) CP   (d) All of above

17. Which of the following mirror formed virtual and enlarged image?

(a) Convex   (b) Concave (c) Plane   (d) All of above

18. The image formed on the focus of a mirror is always:

(a) Real and inverted   (b) Highly Diminished Point size

(c) Virtual and erect   (d) Enlarged

19. The S.I. unit of the focal length is:

(a) Diopter   (b) cm   (c) m (d) m^2

20. Which mirror is used in the rear view mirrors in vehicles?

(a) Plane   (b) Concave (c) both (a) & (b)   (d) Convex

21. The magnification of a plane mirror is:

(a) -1 (b) 1   (c) $\pm$1 (d) All of the above

22. The image distance and the object distance of a mirror is equal. What type of mirror it is:

(a) Concave (b) Plane   (c) Convex   (d) Spherical

23. Which mirror always forms virtual diminished image

(a) Plane   (b) Convex   (c) Concave (d) None of these

24. The relationship between m, v and u is:

(a) m=v/u   (b) mu=-v   (c) –m=u/v   (d) m=uv

25. If the Radius of the mirror is 10cm, object distance = 5cm. Then find the focal length of the mirror.

(a) 2.5cm   (b) 5cm   (c) -2.5cm   (d) 2/5

26. The angle of incident is equal to the angle of ....when light refracted through a glass slab.

   (a) refraction   (b) emergent

   (c) reflection   (d) Angle of the prism

27. When light passing through one medium to other, its direction slightly charge. This cause is known as:

   (a) reflection   (b) refrection   (c) scattering   (d) none of these

28. In Rear medium the speed of the light is:

   (a) slower   (b) faster   (c) same   (d) normal

29. Which is the snell's laws of refraction:

   (a) $<i/<r$ = constant   (b) $<r/<I$ = constant

   (c) $<i/<e$ = constant (d) $<e/<r$ = constant

30. The speed of the light faster in:

   (a) air   (b) Water   (c) Vacuum (d) Solid

31. In which medium light ray bent towards the normal:

   (a) denser   (b) rarer   (c) normal   (d) both (a) & (c)

32. The speed of the light in vacuum is:

   (a) 3×10^8 Km/S   (b) 3×10^8 m/s   (c) 3×10^{-8} m/s   (d) both (a) & (c)

33. Which lens is thicker at the middle as compared to the edges?

   (a) Convex   (b) Concave (c) Plane   (d) bifocal lens

34. Which lens is bulging outwards?

   (a) Convex   (b) Concave (c) Plane   (d) bifocal lens

35. A lens has _____ focus point:

(a) 1   (b) 2   (c) 3   (d) 0

36. The central part of the lens is known as:

(a) Pole   (b) optical centre   (c) radius   (d) centre

37. Which of the following gives a real as well as virtual image?

(a) Convex mirror, concave lens   (b) Convex lens, Concave lens

(c) Concave mirror, convex lens   (d) Concave lens, Convex lens

38. If the power of a lens is -4diopter then what type of lens it is:

(a) Convex   (b) Concave (c) bifocal   (d) both (b) & (c)

39. The Power of a lens is -1diopter, and then find its radius

(a) -1/2   (b) 1   (c) 2   (d) -2

40. The focal length of a Concave lens is 100cm. the find it power

(a) 100 dioptre   (b) -1 dioptre   (c) 1 dioptre (d) -100 dioptre

41. All distances of the lens measure from the _____

(a) Pole   (b) Centre   (c) None of these   (d) optical Centre

42. Which of the following is not behaves like a lens.

(a) water   (b) mirror   (c) glass   (d) both (a) & (b)

43. Which mirror is used in solar furnaces?

(a) Convex   (b) Concave (c) Plane   (d) bifocal

44. Which lens is used in magnifying glass?

(a) Concave (b) Plane   (c) Convex   (d) bifocal

45. The refractive index of water is:

(a) 1.3   (b) 3x10^8 m/s   (c) 1.00033 (d) 1.52

46.   Which of the following is the lens formula?

(a) 1/v-1/u =f   (b) 1/v-1/u=2/R

(c) 1/R-1/u=2v   (d) 1/v-1/u=R/f

47. Refraction can take place in?

(a) convex mirror   (b) glass slab   (c) concave mirror (d) None of these

48.   In a glass slab. The incident ray is _____ to the emergent ray:

(a) Same   (b) equal   (c) parallel   (d) collinear

49.   A Girl saw her in a magic mirror she saw the upper part in same middle is enlarged and the lower is diminished in the mirror. Then what types of mirrors are:

(a) Convex, Concave and Plane   (b) Plane, Convex and Concave

(c) Plane, Concave and Convex   (d) Concave, Plane and Convex

ANSWERS:

QUE.	ANS.	QUE.	ANS.	QUE.	ANS.	QUE.	ANS.	QUE.	ANS.
1	D	11	C	21	B	31	A	41	D
2	D	12	B	22	B	32	B	42	B
3	A	13	C	23	B	33	A	43	B
4	A	14	B	24	B	34	A	44	C
5	B	15	B	25	B	35	B	45	C
6	C	16	B	26	B	36	B	46	B
7	D	17	B	27	B	37	C	47	B
8	B	18	B	28	B	38	B	48	C
9	B	19	C	29	A	39	D	49	C
10	B	20	D	30	C	40	B		

7. HUMAN EYE

SOME IMPORTANT POINTS

- Light enters in the human eye through cornea.
- The eye ball is 2.3 in diameter.
- Iris regulates the size of pupil.
- Pupil regulates the amount of light entering in the eye.
- The crystalline lens focuses the image on the retina.
- The lens is biconvex lens.
- Retina is light sensitive screen.
- Retina generates electric signal and these signals sent via optic nerve.
- The adjustment of focal length of crystalline lens with the help of cilliary muscle called power of accommodation.
- The near point of normal vision is about 25cm.
- The far point of normal vision is infinity.
- Myopia: In this defect eye can see near object clearly but cannot see distant object distinctly.
- The image formed in this defect front of retina.
- Causes of myopia elongation of eye ball :Decrease in focal lens
- This defect is corrected by using a concave lens.
- Hypermeteropia: A person can see distant object distinctly but cannot see near object clearly.
- The image formed behind the retina.
- Causes of hypermeter 1. Eye ball become too smalt 2. Increase in focal length.
- This defect is corrected by using a convex lens.
- Presbyopia: As we become old we cannot near and distant object clearly.
- Causes:
    1. Gradually weakening of a cilliary muscles
    2. Flexibility of eye lens decreases.
- It is correct by bifocal lens upper past consist concave lens and lower past convex lens.

- Prism splits white light into a band of colours.
- These colours are a component of white light is called its spectrum.
- The splitting of white light in to its colour components is called dispersion.
- The red light band least violet band the most.
- In rainbow formation first dispersion and refraction of light takes place and internal after this again dispersion and reflection takes place.
- Twinkling of star apparent position of star advance sunrise and delayed sunset all phenomena are happening due to atmospheric refraction.
- The scattering of light by colloidal particle is called tyndall of effect
- The colour of sun at sunrise and sunset all happens due to scattering of light.

7. THE HUMAN EYE

1. We can identity objects by a sense organ called?

   (a) nose   (b)eye   (c)tongue   (d)none of these

2. The image formed on a light sensitive screen is called?

   (a)retina   (b)cornea   (c)iris   (d)none of these

3. Light enters the eye through a membrane called?

   (a)iris   (b)cornea   (c)retina   (d)none of these

4. The diameter of eyeball is approx?

   (a)2.3cm   (b)2.5cm   (c)3cm   (d)none of these

5. A dark muscular diaphragm which controls the size of pupil is?

   (a)iris   (b)pupil   (c)cornea   (d)none of these

6. The amount of light entering the eye is regulated by?

   (a) iris   (b)pupil   (c)cornea   (d)none of these

7. The adjustment of focal length required for objects at different distances is provided by?

   (a)iris   (b)crystalline lens   (c)eye ball   (d)none of these

8. The image formed on retina is?

   a) real and inverted   (b)virtual   (c)erect   (d)none of these

9. The electrical signals generated by light sensitive calls are sent to brain via?

   (a) Cornea   (b) retina   (c) optic nerve   (d) none of these

10. In the dim light iris expands the?

    (a) Retina   (b) pupil   (c) cornea   (d)none of these

11. The visual impairment is due to damage of?

    (a) Retina or optic nerve (b)iris or pupil

    (c) Cornea or eye lens   (d) all of these

12. We can see distant objects clearly by increasing in?

    (a) Curvature (b) focal length (c) pole (d)none of these

13. We can see nearby objects by increasing in?

    (a) Curvature   (b) focal length (c)pole (d)none of these

14. The ability of eye lens to adjust its focal length is called?

    (a)refraction (b)magnification (c)accommodation (d)none of these

15. The near point of view of a normal eye is?

    (a)20 cm (b)25cm (c)30 cm (d)none of these

16. The far point of view of a normal eye is?

    (a) Infinity (b) 40 cm (c) 25cm (d)none of these

17. The crystalline lens of people milky and cloudy this condition is?

    (a)myopia (b)cataract (c)magnification (d)none of these

18. A person can see nearby object clearly but cannot see distant objects distinct by due to?

    (a)myopia (b)presbyopia (c)hypermetropia (d)none of these

19. The defect which arise due to excessive curvature of the eye lens is

    (a)myopia (b)presbyopia (c)hypermetropia (d)none of these

20. Myopia can be corrected by using?

    (a) Concave lens (b) convex lens (c)bifocal lens (d)none of these

21. A person can see distant objects clearly but cannot see near objects distinctly due to?

    (a)myopia (b)hypermetropia (c)presbyopia (d)none of these

22. The defect which arises due to increasing in focal length is?

    (a)myopia (b)presbyopia (c)hypermetropia (d)none of these

23. In a myopic eye the image of object is formed on?

    (a) retina (b)front of retina (c)behind of retina (d)none of these

24. In a hypermetropic eye the image of object is formed on?

    (a)retina (b)front of retina (c)behind of retina (d)none of these

25. Hypermetropia can be corrected by image of object is formed on?

    (a) retina (b)front of retina (c)behind of retina (d)none of these

26. A bi focal lens consists of?

    (a) convex lens (b)concave lens (c)both (a)and (b) (d)none of these

27. The defect which arises due to weakening of the cilliary muscles is?
    (a)myopia   (b)hypermetropia (c)presbyopia
    (d)none of these

28. The upper portion of a bi –focal lens consist?

    (a) concave lens (b)convex lens

    (c) both (a)and (b) (d)none of these

29. The lower portion of bi –focal lens is consisting of?

    (a)concave lens (b)convex lens (c)both (a)and (b) (c)none of these

30. It is possible to correct refractive defects by using?

(a)contact lens   (b)surgical intervention

(c)both (a)and (b)   (c)none of these

31. in a glass which ray is parallel to the incident ray during refraction?

(a) normal ray (b)refracted ray (c)emergent ray (d)none of these

32. The number of rectangular lateral surfaces in a glass prism is?

(a)four (b)three (c)two (d)none of these

33. The number of triangular bases in a glass prism?

(a)four (b)three (c)two (d)none of these

34. The angle between two lateral surfaces of a prism ?

(a)angle of emergences (b)angle of prism

(c)angle of incidence   (d)none of these

35. When a light ray entering to the denser medium form rarer medium it bends towards?

(a)normal (b)incident ray (c)emergent ray (d)none of these

36. The spectrum of light means?

(a)band of coloured components (b)group of colures
(c)both (a)and (b)   (d)none of these

37. The various colures present in sunlight are?

(a)violet, indigo, blue, green, yellow, orange, red

(b) violet , indigo , brown , green, yellow ,orange, red

(c) violet , indigo , brown ,pink , yellow , orange

(D) none of these

38. The splitting of light into its colored components is called?

(a) scattering (b)dispersion (c)spectrum (d)none of these

39. The colour which bends the least during dispersion of light is?

(a)violet (b)green   (c)red (d)none of these

40. Who was the first to use a glass prism to obtain the spectrum of sunlight?

(a)gallileo (b)Issac Newton (c)Archimedes   (d)none of these

41. Rainbow of light formed due to?

(a)scattering of light   (b)dispersion of light

(c)refraction of light   (d)none of these

42. A rainbow is always formed in the dire of?

(a) infront of the sun   (b)opposite the sun (c)both (a)and (b) (d)none of these

43. The different colours of rainbow reach to observer's eye due to?

(a) dispersion of light   (b total internal reflection

(c) both (a)and (b)   (d)none of these

44. The hotter air present in atmosphere is?

(a) lighter that cooler air (b)heavier than cooler air

(c)both (a) and (b   (d)none of these

45. The twinkling of stars is due to?

   (a) scattering   (b)atmospheric refraction

    (c) dispersion of light   (d)none of these

46. The time difference between sunset and the apparent sunset is about ?

   (a)5 minutes (b)2 minutes (c)10 minutes   (d)none of these

47. The path of a beam of light become visible when it through?

   (a) A true solution   (b) suspension solution

   (c) Colloidal solution   (d)none of these

48. If the size of scattering particles is large enough the scattered light

   may appear?

   (a)red   (b)white   (c)violet   (d)none of these

49. The colour of sky appear blue due to?

   (a)scattering of light   (b)dispersion of light

   (c)atmospheric refraction   (d)none of these

50. The colour which has the longest wavelength is?
   (a) red   (b)violet   (c)green   (d)blue

ANSWERS:

QUES.	ANS.	QUES.	ANS.	QUES.	ANS.	QUES.	ANS.	QUES.	ANS.
1	B	11	D	21	B	31	C	41	B
2	A	12	B	22	C	32	B	42	B
3	B	13	A	23	B	33	C	43	C
4	A	14	C	24	C	34	B	44	A
5	C	15	B	25	B	35	A	45	B
6	B	16	A	26	C	36	C	46	B
7	B	17	B	27	C	37	A	47	C
8	A	18	A	28	A	38	B	48	B
9	C	19	A	29	B	39	C	49	A
10	B	20	A	30	C	40	B	50	A

ELECTRICITY

SOME IMPORTANT POINTS

➢ The rate of flow of electric charge is known as electric current.

➢ Electric charge is denoted by "Q" and its SI unit is coulomb.

➢ When a net charge Q flows in any cross section conductor in time T then current I is I = Q/T (I = current, Q = charge).

➢ The SI unit of current in Ampere.

➢ 1Coulomb in equivalent to charge contained $6.25*10^{18}$ electrons.

➢ 1Electron passes negative charge is usually is nearly $1.6*10^{-19}$ coulomb.

➢ A continuous and closed path of an electric current is called an electric circuit.

➢ Electric current is measured by a device called Ammeter.

➢ It is always connected in series in a circuit.

➢ An ideal Ammeter has low resistance.

➢ The electron moves only if there is a difference of electric pressure called the potential difference along the conductor.

➢ The electric potential difference is defined as difference between two points in an electric circuit carrying some current as the work done to move a unit charge form one point to the other.

➢ Potential difference is denoted by "V" and its SI unit is volt. V= W/Q, 1V = J/C.

➢ Potential difference is measured by an instrument called voltmeter and it is always connected in parallel in a circuit.

➢ It is high resistance.

➢ Ohm`s law stated as electric current is flowing through a circuit is directly proportional to potential difference across it ends provided its temperature remains the same. V α I, V= IR.

➢ Where R is constant and it is called resistance. Resistance is a property that resist the flow of electrons in a conductor and its SI unit is ohm Ω. R= V/I.

➢ A component used in a circuit to regulate the electric current with charging voltage source called rheostat.

➢ Resistance of conductor depends upon :
   1. Length of wire.

2. Area of cross section.
3. The nature of its material.
4. Temperature.

➢ R α l,   R α l/a,   R α l/a, $R = \rho l/a$

➢ Where ρ is proportionality constant called 'rho' and it is called resistivity of materia

The SI unit of resistivity is ohmmeter (Ωm).

➢ Tungsten is used in filament of electric bulbs.

➢ Resistors connected in two ways series and parallel.

➢ In series the equivalent resistance is given by R_{eq} = R1+R2+R3.....

➢ In parallel the equivalent resistance is given by 1/Req = 1/R1 + 1/R2 + 1/R3......

➢ Joule law of heating $H = I^2 RT$, H = VIT where H is electric energy electric power is H/
$P = I^2R$ and $P = V^2/R$, P = VI.

➢ The rate at which electric energy is dissipated or consumed in a electric circuit is cal
power.

➢ The SI unit of power is watt.

➢ The commercial unit of electric energy is kWh commonly known as unit.

➢ 1kwh = 3.6×10^6 Joule.

8. ELECTRICITY

1.   Which quantity is either positive or negative?

   a. Time   b. Charge   c. Volt   d. Both a and b

2.   1C net charge is equivalent to the charged contained in nearly ...... electrons.

   a. $6.25*10^{18}$ b. $6.25*10^{17}$ c. $6.25*10^{-18}$ d. None of these

3.   Which quantity has SI unit coulomb?

   a. Resistance   b. Current   c. Voltage   d. Charge

4.   Rate of flow of net charge is known as?

   a. Resistance   b. Charge   c. Current   d. Power

5.   The SI unit of current is?

   a. Ampere   b. Second   c. Volt   d. Resistance

6.   Current * time =?

   a. Resistance   b. Charge   c. Volt   d. Resistivity

7.   1 micro Ampere =?

   a. Time   b. Charge   c. Volt   d. Both a and b

8.   In an electric circuit the electric current flow in an .... direction to the flow of electron conventionally.

   a. same   b. opposite   c. both a and b   d. None of these

9.   Which quantity is measured by ammeter?

   a. Resistance   b. Power   c. Current   d. None of these

10. Ammeter has …………. resistance.

a. High   b. normal   c. current   d. low

11. In the electric circuit how many cells are there?

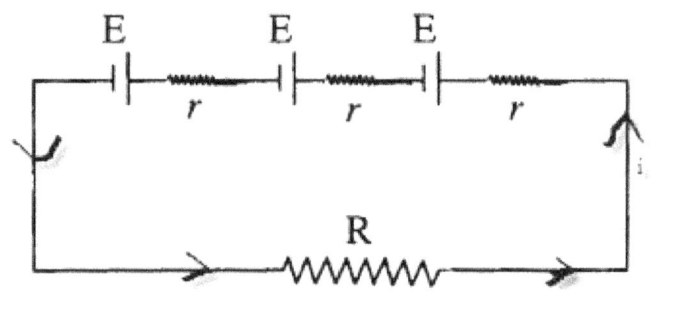
Diagram

a. 2   b. 3   c. 4   d. 5

12. Voltage *Charge =?

a. Current   b. Resistance   c. Work   d. None of these

13. Potential difference is denoted by?

a. W   b. Q   c. V   d. P

14. Which quantity is measured by voltmeter?

a. work   b. Charge   c. Potential difference   d. Power

15. Which of the following is scalar quantity?

a. Current   b. Potential Difference   c. Charge   d. All of these

16. Voltmeter has ……….. resistance.

a. High   b. Normal   c. Both a and B   d. low

17. Which is always connected in parallel across the circuit?

a. Ohmmeter   b. Ammeter   c. Speedometer   d. voltmeter

18. What is the symbol of joint wire?

a.

b.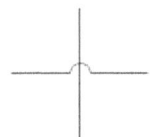

c.

d.   none of these

19.   Who found the relationship between the current (I) flowing through a conductor and potential difference (V) across the terminals of a conductor using circuit diagram.

a. George Simon Ohm   b. Andrew Marie Ampere

c. Glenn Maxwell   d. None of these

20.   Resistance =?

a. Voltage *Current   b. Voltage/Current

c. Current/Voltage   d. All of these

21.   The property of a conductor that opposes the flow of current is known as?

a. Resistance   b. Rheostat   c. Voltage   d. None of these

22.   What is the SI unit of resistance?

a. Volt   b. Ohm   c. Watt   d. None of these

23.   Which of the following is a resistor?

a. Human   b. Wire   c. Wood   d. All of these

24.   The resistance of a conductor depends on?

a. temperature   b. Volume of material c. both a and b   d. None of these

25.   What is the SI unit of resistivity?

a. Kilowatt-hour   b. ohm-meter   c. ohm   d. volt

26.   Which element is used almost exclusively for filament of electric bulb?

a. Titanium   b. Copper   c. Tungsten   d. alluminium

27.   Which elements are generally used for electrical transmission lines?

a. Alluminium   b. Copper   c. Both a and b   d. None of these

28.   Which of the following is a series combination?

a.

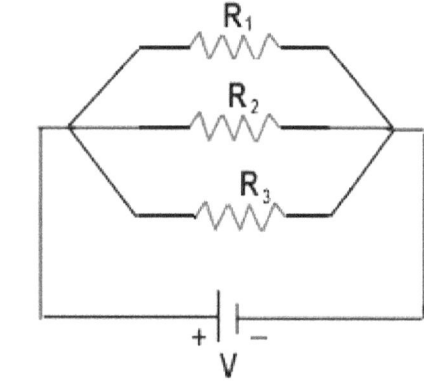

b.

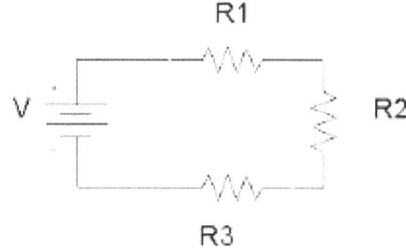

c. Both a and b

d. None of these

29.   Which quantity remains same in combination of resistors?

a. Potential difference   b. Current   c. Charge   d. Time

30. Which gases are filled in bulb to prolong the life of filament?

   a. Nitrogen and Helium   b. Helium and Argon

   c. Argon and Oxygen   d. Nitrogen and Argon

31. 1 Volt * 1 Ampere =?

   a. 1C   b. 1J   c. 1W   d. 1A

32. A continuous and closed path of an electric current is known as?

   a. Electric circuit   b. Electric current   c. Electric rode   d. None of these

33. A current of 0.2 A is drawn by an electric bulb for 22 minutes. Find the amount of electric charge that flow through the circuit.

   a. 2640 A   b. 264 C   c. 264 A   d. 2640 C

34. The filament of an electric CFL draws a current of 0.5A for 2 hours. Calculate the amount of charge that flows into the circuit.

   a. $3.6*10^2$ C   b. $3.6*10^{-2}$ C   c. $3.6*10^3$ C   d. $3.6*10^{-3}$ C

35. Calculate the work done in moving a charge of 12 C across two points having a potential difference of 7 V.

   a. 84 J   b. 84 W   c. 84 C   d. 84 ohm

36. Calculate the work done in moving a charge of 20 C across two points having a potential difference of 5V.

   a. 8*100 J   b. 8^0*100 J   c. 8*1000   d. None of these

37. Given R = 440 ohm, V = 220 V, then I=?

   a. 0.5 A   b. 0.05 A   c. 5 A   d. All of these

38. Given, V = 220 V, R = 600 ohm, I =?

a. 0.003   b. 0.3   c. 0.0003   d. 0.03

39.   Given V = 100 V, I = 5 A, R =?

       a. 20 ohm-m   b. 20 ohm   c. 200 ohm-m   d. 200ohm

40.   3 resistor connected in parallel combination of resistance 10ohm, 20 ohm, 25 ohm. What is its equivalent resistance?

       a. 0.019 ohm   b. 0.19 ohm   c. 100/9 ohm   d. 19/100 ohm

41.   Given P = 840 W, t = 10s, W = ?

       a. 8400 J   b. 8400 C   c. 8.4 J   d. None of these

42.   Given I = 10 A, R = 10 ohm, t = 0.5 s, H = ?

       a. 8400 J   b. 8400 C   c. 8.4 J   d. None of these

43.   An electric bulb is connected to a 220 v battery. The current is 3 A. what is the power of bulb?

       a. 6.6 W   b. 66 W   c. 660 W   d. All of these

44.   A CFL is rated at 5V, 100A. What is its power?

       a. 500 W   b. 5^0*100 W   c. 50*10 W   d. Both a and c

45.   An electrical fan rated 200W operates 4 hours per day. What is the cost of the energy to operate it for 30 days at the rate of Rs.2 per KWh?

       a. 28   b. 40   c. 288   d. None of these

46.   A charge of 700 C flowing in 10 s. How much current is flowing through the circuit?

       a. 70 A   b. 70 J   c. 70 W   d. None of these

47.   Calculate the work done in moving a charge of 120 C across two points having a potential difference of 5 V?

a. 500 J   b. 600 J   c. 525 J   d. None of these

48.   H = 25 J, I = 5A, t = 1s , R = ?

a. 1 ohm   b. 0.1 ohm   c. 0.01 ohm   d. 0.001 ohm

49.   An electrical heater takes 10 A current from a 220 V line. Determine power of heater and energy consumed in 5 h.

a. 2200, $3.96*10^8$ J   b. 2200, $3.96*10^7$ J

c. 2020, $3.96*10^7$ J   d. 2000, $3.96*10^7$ J

ANSWERS:

QUE.	ANS.	QUE.	ANS.	QUE.	ANS.	QUE.	ANS.	QUE.	ANS.
1	D	11	B	21	A	31	C	41	A
2	A	12	C	22	B	32	A	42	B
3	D	13	C	23	D	33	B	43	C
4	C	14	C	24	A	34	C	44	D
5	A	15	D	25	B	35	A	45	D
6	B	16	A	26	C	36	B	46	A
7	C	17	D	27	C	37	A	47	B
8	D	18	A	28	B	38	D	48	A
9	C	19	A	29	B	39	B	49	B
10	D	20	B	30	D	40	C		

9. MAGNETIC EFFECT OF ELECTRIC CURRENT

SOME IMPORTANT POINTS

➤ The wire carrying electric current behaves like a magnetic.
➤ The area in which force of attraction & repulsion is felled called magnetic field
➤ Magnetic field cannot intersect each other
➤ Magnetic field denser at poles.
➤ Magnetic field originates from North Pole to South pole in bas magnet.
➤ Load stone is natural magnet called Hematite.
➤ Magnetic field is rector quantity.
➤ Magnitude of magnetic field is directly proportional to current.
➤ Right hand thumb rule stated as if your thumb to be the direction of current then your wrapped fingers show the direction of magnetic field lines.
➤ In a circular coil having turns the field produced times.
➤ Field lines inside the solenoid is parallel and it behaves like a bar magnet.
➤ Fleming's left hand rule: If the fore finger shows the direction of field lines and the middle finger show direction of current then thumb shows the direction of motion
➤ The principle of electric motor is Fleming's left hand rule and electric motor convert electrical energy in to mechanical energy.
➤ The principle of generator based upon electromagnetic Induction.
➤ The process by which a changing of magnetic field in a conductor induced current is called electromagnetic induction.
➤ In generator mechanical energy converted into electrical energy.
➤ The time varying current is known as alternative current.
➤ The current does not its direction with time called direct current
➤ Ac changes direction after every 1/100 second in India and frequency of AC is 50 H$_z$
➤ In home all appliances connected in parallel.
➤ Fuse is a safety device of all domestic circuits.

- Fuse is made up of alloy of copper and tin.
- In our houses we get AC of 220 volt.

9. MAGNETIC EFFECT OF ELECTRIC CURRENT

1. An electric current – carrying conductor wire behaves like a?

    a. magnet   b. conductor   c. galvanometer d. motor

2. The electric current through the copper wire has produced a:

    a. magnetic effect   b. condutivity   c. tindal effect   d. none of these

3. Magnet has properties of……..and………

    a. attraction b. both (a) and (c)   c. repulsion   d.none of these

4. Magnets consist of a number of oxides of iron with……… its structure

    formula:

    a. Fe_3O_4 b. co_2 c.Fe_4O_3 d.FE_4O_3

5. Natural magnets are:

    a.Strong   b. weak   c.   irregular In shape   d.both (a)and (b)

6. The chemical properties of magnet and iron is:

    a.opposite   b.same   c.same but different   d. none of these

7. A compass needle gets deflects when we brought it……..

    a.near a bar megnet   b.far from a bar megnet

    c. carry a bar megnet   d. none of these

8. A bar magnet is a:

    a. natural magnet   b.electromagnet

     c. artificial magnet   d. none of these

9. There are two ends of magnets:

    a. south- north   b. southern east- north

c. northern east- west   c. east-west

10. When an iron and magnet comes into contact:

   a. attract each other   b. repel each other

   d. crunch each other   d. all of above

11. ....poles of magnet attract each other.

   a. unlike   b. both (a)and (c) c. like   d. none of these

12. Choose the incorrect statement:

   a. magnet field lines are parallel and equidistant.

   b. Fe_3O_4 is the structural formula of magnet.

   c. magnetic field lines are closed and curve.

   d. both a and c.

13. What is the SI unit of magnetic field line?

   a. ampere   b. torque   c. Newton   d. tesla

14. The Fleming's left hand rule applies on?

   a. motor   b. generator   c. T.V   d. Bike

15. choose the correct option:

   a. the magnet exerts its influence in its surroundings

   b. magnetic field lines form closed loop.

   c. in presence of magnetic field the iron fillings are arranged in a pattern.

   d. all of these

16. When current flows magnetic field produced is .... to the direction of the current flow.

a. parallel   b. perpendicular   c. straight   d. none of these

17. When the magnitude of the current increases the magnitude of the magnetic field ………..?

a. increases   b. decreases   c. first increases then decreases   d. None

18. If magnitude of magnetic field increases then deflection in compass…..?

a. increases   b. decreases   c. remains constant   d. both a and c

19. The magnetic field lines emerges from?

a. south pole   b. north pole   c. either south or north   d. none

20. The magnetic field lines merge at?

a. north pole   b. south pole   c. either south or north   d. none

21. Inside the magnet the magnetic field direction is from?

a. north to south   b. north to east   c. south to east   d. south to north

22. Magnetic field is?

a. vector   b. scalar   c. Constant   d. none

23. The field lines are stronger when we keep the magnets.

a. closer   b. far   c. field lines are same everywhere   d. none

24. Choose the correct option.

a. magnetic field lines merges at south pole

b. magnetic field lines are invisible

c. like pole repel each other

d. all of these

25. Choose the correct option.

a. no two field lines cross each other

b. Two field lines crosses each other

c. magnetic field lines are parallel to each other

d. none of these

26. An electric current through a metallic conductor produces:

a. electric field   b. cooling   c. energy   d. magnetic field

27. If the current flows from north to south the north pole of the compass needle would move towards the?

a. west   b. east   c. north   d. south

28. The compass deflects due to:

a. magnetic field   b. electric field   c. field lines   d. none

29. In which element electric field lines surround it?

a. conductor   b. insulator   c. semiconductor   d. none

30. Magnetic field lines are …………..to the electric field.

a. parallel   b. perpendicular   c. intersecting   d. none

31. Magnetic field lines are generated by current through a:

a. conductor   b. insulator   c. semiconductor   d. none

32. If we reverse the direction of electric current, the direction of magnetic field will be?

a. reverse   b. forward   c. no change   d. none

33. If current increases:

a. deflection in compass increases

b. deflection in compass decreases

c. remains constant

d. none

34. Right hand thumb rule indicates:

a. magnetic field lines are parallel to electric current

b. magnetic field lines are perpendicular to electric current

c. magnetic field lines are bisectors of electric field lines

d. none

35. The magnetic field produced by current carrying straight wire depends on the distance from it, as

a. inversely   b. directly   c. 4 times   d. none

36. By time we reach the centre of the circular loop, the arc of the big circle would appear as:

a. straight line   b. curved line   c. intersecting line   d. none

37. To check whether the every section of the wire contribute to the magnetic field lines in the same direction within the loop, we apply:

a. right hand rule   b. left hand rule   c. both a and b   d. none

38. Every point on the wire carrying current would give rise to the magnetic field appearing as straight line at the centre of the:

a. loop   b. motor   c. tube light   d. generator

39. Loop is a:

a. conductor   b. semiconductor   c. insulator   d. None

40. Magnetic field is the:

a. force   b. energy   c. both a and b   d. none

41. A coil produces magnetic field:

a. when electrons flow   b. when electrons becomes static

c. due to magnetic force of coil   d. none

42. The current in each circular turn has the;

a. same direction   b. opposite direction

c. parallel direction of coil   d. none

43. A coil of many circular turns of insulated copper wire wrapped closely in the shape of cylinder is called a:

a. loop   b. coil   c. none of these   d. solenoid

44. The pattern of magnetic field lines of the solenoid is same as pattern of:

a. bar magnet   b. conductor   c. loop   d. coil

45. One end of solenoid behaves as North Pole; second end of the solenoid behaves as:

a. south pole   b. north pole   c. east pole   d. none

46. Soft iron when placed inside the coil then it changed to:

a. electromagnet   b. conductor   c. plastics   d. None of these

47. The magnetic field inside a long solenoid carrying current:

a. is zero   b. decreases as we move towards its ends

c. increases as we move towards its ends   d. is same at all points

48. The magnet must exert an equal and opposite force on current carrying:

a. conductor   b. insulator   c. semiconductor   d. None of these

49. The device used for producing electric current is:

a. Galvanometer   b. Remote   c. Generator   d. Motor

50. The direction of the force on the conductor depends upon the direction of:

a. current and magnetic field   b. magnetic and electric field

c. current and force   d. None of these

51. An electric motor converts electrical energy into:

a. Mechanical energy   b. potential energy   c. kinetic energy   d. none

52. The energy stored in cell which produces electrical energy is:

a. south pole   b. north pole   c. east pole   d. none

53. Flow of charge due to varying magnetic field with respect to the conductor is called:

a. magnetic flux   b. electromagnetic induction

c. magnetic field   d. None of these

54. The direction of the magnetic field is perpendicular to the current motion. It experience .... for moving a coil.

a. pressure   b. thrust   c. force   d. none

55. In an electric generator the .... is used to rotate a conductor in a magnetic field to produce electricity.

a. mechanical energy   b. potential energy   c. kinetic energy   d. none

ANSWERS:

QUE.	ANS.	QUE.	ANS.	QUE.	ANS.	QUE.	ANS.	QUE.	ANS.	QUE.	ANS.
1	A	11	A	21	D	31	A	41	A	51	A
2	A	12	A	22	A	32	A	42	A	52	D
3	B	13	D	23	A	33	A	43	D	53	B
4	A	14	A	24	D	34	B	44	A	54	C
5	D	15	D	25	A	35	A	45	A	55	A
6	A	16	B	26	D	36	A	46	A		
7	A	17	A	27	B	37	A	47	D		
8	A	18	A	28	A	38	A	48	A		
9	A	19	B	29	A	39	A	49	C		
10	A	20	A	30	B	40	A	50	A		

NOTES